POPO AND FIFINA

*The whole family came out in front of the
thatched hut.*

POPO
and
FIFINA

Arna Bontemps
and
Langston Hughes

Illustrations by
E. Simms Campbell

Introduction and Afterword by
Arnold Rampersad

OXFORD UNIVERSITY PRESS
NEW YORK · OXFORD

Oxford University Press

Oxford   New York   Toronto
Delhi   Bombay   Calcutta   Madras   Karachi
Kuala Lumpur   Singapore   Hong Kong   Tokyo
Nairobi   Dar es Salaam   Cape Town
Melbourne   Auckland   Madrid
and associated companies in
Berlin   Ibadan

Published by Oxford University Press, Inc.,
200 Madison Avenue, New York, New York 10016

Oxford is a registered trademark of Oxford University Press

Library of Congress Cataloging-in-Publication Data
Bontemps, Arna Wendell, 1902-1973.
Popo and Fifina / Arna Bontemps and Langston Hughes :
illustrations by E. Simms Campbell :
introduction and afterword by Arnold Rampersad.
p. cm. — (The Iona and Peter Opie Library)
Summary: Popo and Fifina move from the country to a village in Haiti
where papa Jean plans to earn a living as a fisherman.
ISBN 0-19-508765-8
[1. Family life—Fiction. 2. Moving, Household—Fiction.
3. Blacks—Haiti—Fiction. 4. Haiti—Fiction.]
I. Hughes, Langston, 1902-1967.
II. Campbell, E. Simms (Elmer Simms), 1906-1971, ill.
III. Title.   IV. Series.
Pz7. B6443Po   1993   38841
[Fic]—dc20   93-8457
CIP AC

1 3 5 7 9 8 6 4 2
Printed in the United States of America
on acid-free paper

CONTENTS

Popo and Fifina, which was first published in 1932, is the result of the work of two accomplished writers, Langston Hughes and Arna Bontemps. Although they went on to write many other books for children, this was the first children's book published by either author. Eight years before, in 1924, Hughes and Bontemps had met for the first time in New York City, to which each had come in search of fame and fortune as a writer. They became close friends. In fact, they looked so much like one another that each was sometimes mistaken for the other. They also proved successful as writers. Before writing *Popo and Fifina,* Hughes published two books of poetry and a novel, and Bontemps had published a novel. In 1932, they decided to write a children's book because they loved both books and children, and they also wanted to make the most of their friendship.

In certain ways, the two young men (Hughes was about nine months older) were different. Unlike

Hughes, who never married, Bontemps was both married and a father when they started to write *Popo and Fifina*. Bontemps had never left the United States, but Hughes loved to travel and had already visited Africa and the Caribbean and had lived in Europe. However, the men had certain qualities in common. Both were poets at heart. They were also deeply interested in the history and culture of African Americans, even as they were eager to write for everyone because they believed in fostering friendship and respect among the different peoples of the world.

They decided on a story set in the Caribbean nation of Haiti, where Hughes had spent three months the previous year. In fact, he had passed most of this time in the very community in the north of the country that is the setting of *Popo and Fifina*—Cape Haiti, or Cap-Haitien, as Haitians call it. While there, Hughes had made it his business to get to know as many people as possible. He made friends not only with important and prosperous individuals but also with poor people, of whom there were many in Haiti. He observed the women washing their clothes in the streams and men playing cards in the town. He played games with their children. He made friends with fishermen, who took him out in their boats. He talked to vendors in the marketplace, and he was allowed to attend some of the religious ceremonies, accompanied by drumming, that most foreigners never saw.

Whether Hughes and Bontemps based the characters of Popo and Fifina and their family on actual people is impossible to know. However, Hughes had met many little boys and girls like the two children at the center of their story. Perhaps Arna Bontemps, with his growing family, drew on his personal experience to help with the development of the story. In any event, he and Hughes sat down together, decided on the characters they would write about and the outline of their tale, then perhaps took turns writing and rewriting different parts of the story. Whatever their arrangement, *Popo and Fifina* is a seamless narrative.

In this story, Hughes and Bontemps wanted to tell readers in the United States about the people of another country, Haiti. Most of these people were of African descent, as were the two writers themselves. Hughes and Bontemps wanted to show how the poor people of Haiti actually lived. They sought to show them as lacking almost all of the comforts that citizens of the United States take for granted but nevertheless as loving, proud, hardworking, and disciplined.

Popo and Fifina tells of two poor but happy children, their close-knit family, and the honest way of life they follow in spite of their poverty. Through the artistry of Hughes and Bontemps, we see the beauty of the land and its people in a narrative that is itself lyrical, lovely, and finally unforgettable.

POPO AND FIFINA

I

GOING TO TOWN

POPO and Fifina were walking barefooted behind two long-eared burros down the highroad to the little seacoast town of Cape Haiti. Bags, woven of grass, were hung across the backs of the pack animals, and in these were all the belongings of their family. Popo and Fifina were moving. They were moving from their grandmother's home in the country to a new home in town.

Their parents, Papa Jean and Mamma Anna, were peasant farmers. But they had grown tired of the life on their lonely hillside, and they were going to Cape Haiti, where Papa Jean planned to become a fisherman.

The sunshine was like gold. The little dusty white road curved ribbonlike among the many hills. It was overhung with the leaves of tropical trees. On the warm countryside there was no sound but the droning of insects and the sudden crying of bright birds. There was no hurry or excitement. And the burros' lazy steps set the speed for the little band of travelers.

From the rear of their tiny caravan, Popo could see the members of his family stretched along the road, one after another, like a line of ducks. First was Papa Jean himself, a big powerful black man with the back torn out of his shirt. He wore a broad turned-up straw hat and a pair of soiled white trousers; but, like all peasants of Haiti, he was barefooted. He walked proudly, and there was a happy bounce in his step as he led his little family toward the town of his dreams.

Next came Mamma Anna with the baby, Pensia, swinging from her side as Haitian babies do. Mamma Anna was also barefooted, and she wore a simple peasant dress and a bright peasant headcloth of red and green. She was a strong woman with high glossy cheek bones. She followed her husband step for step in the dusty road. The two loaded burros came next. They were shaggy animals, and their heads were lowered as they trudged along. Great bright-

winged flies rode on their flanks and buzzed around their heads.

Fifina, Popo's ten-year-old sister, walked so close behind the second burro that she could reach out her hand and smack him on the flanks if he stopped to nibble grass on the roadside. She wore a little blue dress that reached her knees.

But Popo, who walked behind her, was only eight, and all he wore was a shirt that didn't even reach his waist. At home he wouldn't have been wearing this; but when a person is making a long journey to an important town like Cape Haiti, he has to dress up a little. So Popo had worn his Sunday clothes. And Sunday clothes for black peasant boys in Haiti usually consist of nothing more than the single shirt Popo was wearing to town.

Like all dressed-up people, Popo was proud of himself this afternoon. He was proud also to be going to town to live by the ocean and to see new wonders. So while the little procession swung slowly along, he frisked about like a young colt, stamping the dust and kicking up his heels. Late in the day, near their journey's end, the road led up a hill and they passed between thickets of dense foliage. Beautiful flowering trees were plentiful, their blossoms red like fire, or white as milk.

Looking back, Popo could see forests of palms, mangoes, banana trees, and coffee bushes bordering the road; but he spent very little time in looking back. His mind was set on reaching the top of the hill. He wanted to get a glimpse of the town that was to be their home, and the great ocean from which Papa Jean would fish food and make a living for his family.

Papa Jean was perspiring as he led the way up the mountain. His naked back flashed like metal in the sun.

"Do you think we're almost to the top?" Popo asked Fifina.

"It isn't far," she said. "But you won't help us to get there any sooner by frisking around as you do."

"I can hardly wait, Fifina! If Papa Jean would let me, I could run ahead and be at the top in a minute. Why do you suppose he makes me stay back here, Fifina?"

"Why, it's plain as anything. It's because he knows you'd run so far ahead we could never catch you."

Popo became silent. He couldn't understand the ways of old people. And his thoughts made him sad. "Oh, my!" he said to himself. Then he settled back into the slow gait of the others and made up his mind to be patient till they reached the hilltop.

Presently Papa Jean reached the high point and stood with his hands on his hips. A few moments later Mamma Anna reached his side. Popo could see their backs against the sky— Papa Jean and Mamma Anna standing on the top of the world at the place where the mountain touches the sky. They were in the middle of the road, and soon the faithful burros were there also. One stopped beside Popo's mother and the other beside his father. Then Fifina reached the top and stood beside one of the burros; but before she got there Popo forgot his place, ran ahead of her, and took his stand beside the other animal.

"Well, here we are," Papa Jean said.

"H'm," Mamma Anna hummed without parting her lips. "H'm." She was pleased.

"How fine!" Fifina exclaimed. "Oh, how fine the ocean is! And what a big town!"

Really the town was small, but, compared with the villages Fifina had seen, it was quite impressive.

Popo said nothing. He was too excited to speak, but his eyes swept the whole bright scene below.

This is what he saw: rows of small white houses and buildings that stretched along the curved water front almost as far as his eyes could reach, old ruins and battlements overlooking the water at several places, trees growing right down to the water's edge, sailboats in the harbor moving under a slow steady wind, others being sculled by half-naked men, and a number of tiny rocking boats anchored along the beach.

"We've got to keep moving," Papa Jean said, after a long pause. "We've got to find a house in town and get unpacked before night. Then, too, we'll need a bed."

"H'm," Mamma Anna agreed.

Fifina clapped her hands in excitement, and Popo danced with glee as the party of travelers took their former positions in the road and started down the hillside to their new home.

II

WORK TO DO

"CHILDREN," said Papa Jean, when they had found their house, "I want you to make yourselves useful this afternoon. It is getting late, and mamma will have many things to do before night. She will need your help. Can I depend on you?"

"Oh, yes, Papa Jean," Popo and Fifina assured him.

"Well, don't forget. I'm going down to the beach to talk with the fishermen when they come in with their boats."

Fifina rushed immediately into the house where her mother was working, but Popo didn't move. He was an obedient boy, but he regretted almost at once his promise to Papa Jean. Popo was anxious to follow his father down to the beach, to wander about and explore the neighborhood into which they had moved.

There was a tiny rum shop near the beach, and there a great many sailors and wandering men loitered. Popo heard their loud, heavy voices. He was eager to stop and hear what they said. One old fellow, a native of Santo Domingo, had a beautiful big parrot that sat on his shoulder. Popo thought it wonderful that a bird should sit on a man's shoulder; and he thought that bird with its large yellow beak, and its green and red and yellow feathers was the finest bird he had ever seen. What was more wonderful, the bird could talk. Popo wanted to stay longer in front of the rum shop, but there would be no time for that to-day; he would have to content himself with what he could see from his own yard. Maybe, if he were good, Papa Jean would let him run down to the beach another day, and wander through the streets of rickety houses even up into the main part of town.

"Run along," Papa Jean called back. "You mustn't let Fifina do all the helping."

Popo walked away slowly. Before he reached the door, he turned and saw his father walking with his hands in his pockets, smoking a cigar. Papa Jean was certainly a big strong man, but Popo could see that he was as eager to get to the beach as any child could be.

The little house to which Popo and Fifina had come with their parents was just a one-roomed shack with a tin roof. It had no windows, and only one door that was as rough and awkward as the door of a woodshed. There were no steps to the house, for the floor was laid so near the ground that it was easy to walk directly in from the outside.

The yard was big. It contained a large mango tree and two banana plants. And next door there was a very tall palm.

A beautiful big parrot sat on his shoulder.

When Popo went inside he found Mamma Anna and Fifina making a bed for Baby Pensia. They had already set up the big bed in which Mamma Anna and Papa Jean were to sleep, and they had fixed a straw pallet on the floor for Popo and Fifina. Now they were stuffing a gunny sack to make a bed for Pensia.

"What can I do?" Popo asked. "I want to help too."

"The beds are about finished," Mamma Anna said. "But we'll need some dry leaves to start the charcoal fire. You can get those for us."

That pleased Popo. He scrambled out of the house and ran across a sandy slope to a place where the brush seemed thickest and driest. Under the dead thickets he found leaves so dry and parched that they crumpled in his hands. That was the kind he wanted. He could tell

— 11 —

when leaves were good for starting fires: he had gathered them before. So he quickly raked up a little pile with his fingers and took them into his arms.

When he returned, Mamma Anna and Fifina were in the yard. Mamma Anna had unpacked her little charcoal stove, and Fifina stood holding the kettle in which the food would be cooked.

The sun was sliding down the sky fast now. There would not be time to cook a kettle of beans and rice, the usual peasant dish in Haiti, so Mamma Anna simply boiled a few plantains —the huge Haitian banana that went with every meal. Then she warmed over a pot of meat they had brought from their old home.

While the fire burned and the pot sizzled, Mamma Anna squatted, resting her elbows on her knees. When she thought the food was warm enough, she filled Popo's and Fifina's small plates, which they carried a few yards away and placed on the ground.

Papa Jean was still down on the beach. Popo could see him standing beside the little boats, examining the catches of the fishermen, making motions with his hands as he talked, and evidently asking many questions. After a while, the fishermen started up a path to the road, and Papa Jean returned to his house.

Popo finished his plate and asked for more. Mamma Anna explained that she had no plantains left, but that she was about to boil some yams of which he might have a helping when they were done.

In the meantime Papa Jean reached the house with several fine fish that he had procured from the fishermen. Mamma Anna began at once to clean these. And when the yams were done she put the fish in the kettle.

If Mamma Anna had lived in the United States, she probably would have cooked her entire meal before she allowed her children to begin eating. But her stove was very small, and so she cooked one thing at a time—in no particular order. And her children ate what she cooked as soon as it was ready.

"I have arranged to go out with the other men," Papa Jean said, as he sat down to eat. "We must be on the beach before sunrise to raise our sails and leave the harbor by the first wind. In deep water we'll drop our nets. Then when the land breeze comes up in the afternoon we'll sail home with our catch. I'll get my share of fish and peddle them in the market—and bring some home to eat, too. It'll be a good way to begin life in this new town."

"H'm," his wife agreed. "And some day maybe you'll get a boat of your own?"

"Yes. I'll get a boat of my own. You'll help me to weave reed nets down on the beach. At night we'll hang them on that big tree to dry."

Papa Jean pointed to a huge rugged banyan tree growing near the water's edge. It had great gnarled roots that came out of the earth like immense serpents and curled up on the ground.

The nets of many fishermen were drying upon its branches now.

"May I go with you then, Papa Jean?" asked Popo. "When you have your own boat, may I go out with you some time?"

"Of course," said Papa Jean. "You'll go with me many times. And some day we may even take Fifina and Mamma Anna and Pensia just for the sail. Who knows?"

"But it won't be easy to get a boat," said Mamma Anna. "We'll have to work hard."

"Yes," said Papa Jean. "We'll have to work and work and work."

"I am willing," Fifina said. "I am willing to help all I can."

"I too," said Popo eagerly.

"Well, the first thing you'll have to do will be to carry these pots and dishes to the fountain at the corner to be washed. Then, when you return, bring a kettle of fresh water. By that time it will be bedtime. We must rise early in the morning."

Popo and Fifina began collecting the few dishes and stacking them in the kettle. When they were finished they took the heavy kettle between them, each holding the handle, and started down the road to the public fountain. And Popo carried the hollowed-out shell of a round gourd to fill with drinking water.

On the way they stopped to look into the rum shop near the beach. It was a little out of their way, but Popo was so eager to see it again that he persuaded Fifina to stop a moment with him. The soft blue twilight was descending in the street like mist. Suddenly the lights of the rum shop came on. Music began playing. Popo and Fifina heard the happy voices of the sailors and wandering men; they heard their mugs and glasses rattling on the little wooden tables. The man with the parrot was still there, his parrot still talking in a strange language that Popo could not understand.

The street was full of happy noises and voices like music. It did not get dark soon; the twilight lingered. Popo was used to long blue twilights like that, but it was a new thing to hear the exciting voices of the city streets.

"Oh, I think the city is grand," he said to Fifina.

"Yes," she agreed. "I'm certainly glad we came. We must thank Papa Jean again for bringing us."

III

RUNNING WATER

THE next day, in the heat of the morning, Popo lay in the doorway naked. There was no reason why he should wear his little dress-up shirt now, so he rolled in the dirt happily, and without fear of soiling a garment. He felt more comfortable than he had been since the beginning of their journey two days before.

At that early hour there was very little wind stirring, and there was almost no activity on the streets. The trees of the yard were still, their leaves powdered by white dust. The two burros were sleeping in their places at the back of the yard, and the usual tropical flies were buzzing their bright wings about the animals' heads.

Presently Fifina and Mamma Anna came to the door with their arms full of soiled clothes.

"We're going to wash to-day," Mamma Anna said to Popo. "You come with us, if you like."

Popo didn't really feel like moving at the moment. He was so comfortable on the warm ground he felt that he might have stayed there the rest of the day. But he managed to draw himself up and to roll his eyes at Mamma Anna's suggestion. After all, it might not be a bad idea to follow Mamma Anna and Fifina to the washing place. He might have a chance to play in the water. That *would* be a treat. Popo sprang to his feet.

"Yes, indeed, mamma," he exclaimed. "I'd like to go very much. But where will you wash the things—at the fountain at the next corner?"

"Oh, no, son, not there. The fountain is all right for washing dishes or milk cans, or even for bathing babies like Pensia, but it is better to wash clothes in the stream that runs along the street." Then Mamma Anna put the clothes she was carrying in Popo's arms. "Here. You carry these. I'll go back and get Pensia."

Pensia, like Popo, wore no clothes at all, only a bead on a cord around her neck. She was a quiet and well behaved baby, and she seemed as delighted as any one to be going out for the morning work.

The stream along the street was no bigger than the stream in some gutters after a rain. But it was clean sparkling water from the mountain springs flowing in a little stone gully for the convenience of people who did not have private wells in their houses.

While Mamma Anna and Fifina washed the pieces of clothing, one by one, Popo and Baby Pensia played in the water. Popo took good care of his baby sister and seated her on the edge of the little stream where her feet could reach the water while he ran up and down in the middle

— 17 —

of the stream splashing in every direction and
having the time of his life.

Soon Popo noticed other women coming with
their clothes to the little streams. They took their
places farther up the street and began beating
their garments in the water in the same way
Mamma Anna washed. They took each garment,

dampened and soaped it, then put it on the rocky
edge of the gully and gave it a good pounding
with a wooden stick while the soapy water ran
out of it. Before long there was a line of busy
women that reached almost the whole length
of the street. Other youngsters like Popo were
playing in the water, and other babies like Pensia were sitting with their feet in the stream.

A long line of busy women washing clothes.

Occasionally a dog or a goat came to the stream to drink. The day that had seemed so dull and quiet a little while earlier was now full of sounds and movements.

There were by now, too, many people passing along the streets with bundles on their heads. Among them was one youngster who attracted the attention of Popo and Fifina—a little smiling black girl who carried a large wooden tray on her head and a small folding stool on one arm. She carried her burden lightly and happily, as if she had been used to balancing things on her head a long time.

All Haitian youngsters learn to carry burdens on their heads. Popo already knew the trick. He could go to the store for a bar of soap or a basket of fruit and bring it home on his head just as expertly as the little girl was carrying her tray. That was a fine thing for a playful boy like Popo, since he could forget the burden on his head and at the same time have his hands free to play.

But the little girl with the wooden tray and the broad smile was not out to play. She was on a business errand, and Popo could see that her tray was loaded with things to sell. When she was near enough, she unfolded her stool, put it on the ground, and set the tray upon it.

Popo's eyes popped, and his mouth began to water, for he was looking at a great collection of stick candy, large sugary peppermint sticks of pink and white. It was a soft crumbly kind of stick candy; it had a fine peppermint smell, and

when the little girl removed the tray from her head she had to shoo the tropical bees and flies away. A whole swarm of them had been sitting happily on the sweet sticks as they traveled along the road uncovered in the sunshine.

"Will you have a stick of candy?" the little girl asked. "It's a penny a stick and awfully good. My mother just made it. Will you buy a stick?"

Popo turned to his mother with a pleading glance, but she was shaking her finger in the air. Fifina looked up eagerly. But pennies are scarce in Haiti. And Mamma Anna was not at all sure that she could afford to spend two or three of them that day for candy.

"Please, mamma," Popo said softly.

"We haven't had a taste of candy for months," Fifina begged.

The mother paused, thinking.

"I don't know," she said. "Not now anyhow. But maybe when the morning's work is done, when the little girl passes here on her way home, you may each have a stick."

Popo clapped his hands and gave an excited leap in the water. Fifina showed her happiness by bringing the back of her hand across her open mouth in a little gesture. Then the girl gathered up her wares, folded her chair across her arm, and started down the street again. Fifina and Mamma Anna returned to their clothes, and once again Popo splashed and galloped in the water. Baby Pensia, who kicked her feet, gurgling, seemed also to understand.

When the washing was done, Mamma Anna wrung the clothes as dry as she could and stacked them in a large tin pan. The loaded pan she lifted quickly to her head. Then, leaving Fifina and Popo to attend to Pensia, she started off toward home, a great pile of whiteness balanced on her head.

At home Mamma Anna unfolded her clothes and spread them carefully on the grass around the house. There was no such thing as a clothes line in anybody's yard.

"Will there be anything for me to do?" Popo asked.

"Oh, yes," Mamma Anna told him. "I will need you to go down by the roadside and get me some soap weed to wash dishes. I used the last bit of bar soap on the clothes, and if you are to have your stick candy, I shan't be able to spend pennies for another bar. You will have to get me a good supply of soap weed."

"I'll go too," Fifina offered. "I know a soap bush better than he does."

The two youngsters went running down the path. And sure enough Fifina led Popo to a large clump beside the roadside. They tore off a few leaves to try it out. Rubbing the leaves between the hands produced a lather not unlike that from moistened soap.

"These will do," Fifina said. "Let's gather as many leaves as we can carry in our hands."

"All right," Popo went to work eagerly.

Ten or fifteen minutes later, when Popo and Fifina and Mamma Anna were at the fountain and Baby Pensia, getting a real bath, was covered with white suds that looked like wool, the little candy girl returned. Her tray was not empty, but it was plain that she had made sales since passing the family at the washing place. True to her word, Mamma Anna took two pennies from a pocket of her skirt and bought one stick for Fifina and one for Popo. Fifina's was white and Popo's was pink.

"Pensia can have a taste of each," she explained. "She does not need a whole stick."

"Glook!" said Pensia, crowing at the sight of the candy. "O—oo! Glook!"

IV

BY THE SEA

ONE afternoon, when there was no work to do and the day was bright with golden sun, Popo and Fifina went down to the beach.

Behind their house there was a gentle slope of about one hundred yards. At the end of the slope there was the large tree with the gnarled serpentlike roots curled above the ground. And a few feet beyond, along the water's edge, were the large rocks of the wave line.

Looking up and down the long curved coast, Popo could see that the harbor of Cape Haiti was shaped roughly like a horseshoe. He could see that almost all the way around mountains rose sharply out of the sea, rocks jutted out of the water itself, and almond trees grew among the rocks.

Popo and Fifina sat side by side on a large rock and looked out across the bay. Away out, some big steamships were anchored, and with them there were a large number of sailing boats.

"Aren't they fine!" Popo exclaimed.

"They are," Fifina agreed.

"But look at these tiny little boats pulling away from the shore. What are they?"

"They are sculling boats," Fifina said. "Those things behind them that the men wiggle like tails are what Papa Jean calls sculls. They are as good as oars, he says."

Popo was looking at a little craft no longer than a good-sized skiff. Three half-naked black men, standing at the end of the boat, were working the sculls back and forth, back and forth, very leisurely, very much indeed like tails. And somehow the motion of these tails sent the boat forward.

"Look," Popo said. "Look at that one near the shore. The men are wading in the water and pushing it."

"Yes," said Fifina. "But do you see what they are carrying in the boat?"

Sitting in the bow of the small craft was a boy about Popo's size. He was naked, and he held under each arm a game chicken. In addition to the boy with the chickens, the boat held a basket of mangoes, two bunches of bananas, and a tiny green parrot tied by the foot and sitting on one of the sculls.

Popo jumped to his feet and threw up his hands, waving at the other youngster. When he saw that the boy was looking at him, he called at the top of his voice, "Say, where are you going with the chickens?"

The boy smiled broadly.

"We are going out to the ships to sell them," he shouted back.

The men who had been pushing the boat out into the deeper water jumped aboard and began working the sculls; and promptly the little bark with its curious cargo drifted out into the blue bay.

Popo stretched out on the rock, rested his chin in his hand, and began daydreaming. He wondered what kind of boat Papa Jean had gone out in, and where he might be at that very moment. Was he selling things to the steamers anchored near the horizon? Or was he out beyond the harbor on the big tossing waves with his net cast in the deep water? Either of these seemed to Popo a fine occupation, and he longed with all his heart to be with Papa Jean. But some day they would have a boat of their own, Papa Jean had promised that, and then he would go out like the youngster with the chickens. Ah, wouldn't that be a life!

Meanwhile Fifina was hopping from rock to rock. Sometimes she stopped to look down into the shallow clear water. She would stand very still for a moment or two, and then she would start leaping and climbing again. Suddenly she came to a quick stop and called very loudly: "Popo! Oh, Popo, come here quick!"

Popo did not wait to ask what she wanted but jumped up and ran around to the rock where she was standing. The rock was under an almond tree that hung over the water, and it was so far out in the water that it could be reached only by stepping on another rock and making a little jump.

"What is it? What is it?" Popo whispered breathlessly as he stood by his sister's side.

"There. See." She pointed to the water near the base of the rock.

"Oh, yes!"

Popo slid down on his stomach, his head hanging over the edge of the rock. Fifina knelt beside him, supporting herself by her hands as she peered into the clear transparent water.

Down near the white sandy bottom they could see a host of lovely red, blue, and yellow parrot fish darting about excitedly. They looked as bright and pretty as sticks of candy, and neither Popo nor Fifina had ever seen any living creatures half so vivid.

"Do you think we could catch some of them?" Popo asked eagerly.

Fifina shook her head.

"We have nothing to catch them with," she said. "Besides, what would we do with them if we did?"

"We might have Mamma Anna cook them for our supper."

"It would be a shame to eat them," Fifina said. "They are such darlings."

Popo thought a moment and then calmly agreed. They were too pretty to eat. And maybe it would be just as well not to frighten them.

"You are right," he whispered.

"But we might catch a few crabs and carry them home with us," she suggested. "Mamma Anna loves them."

They climbed back over the rocks and came down to the water's edge at another place and began digging in the sand and scratching in the water with switches.

"Here, Fifina," Popo called presently. "I have the first one. And a beauty he is too. Just look."

He held up a large greenish-red crab, holding it carefully so as not to be snapped by its claws.

"Well, put a rock on him till I get one," Fifina said. "Then we can fasten their legs together so that they can't crawl away."

A few moments later, she caught a crab of her own, and they attached the claws together. Then they continued their search, digging separately. Soon the string of crabs was nearly a yard long, and Fifina suggested that they had enough for one day.

But before they started for home a boat slid up near the bank and Papa Jean with several other men got out and waded ashore. At first Popo was surprised. He had not supposed that it was time for boats to be returning. But here indeed was his father, and out in the bay were many other small craft, taking advantage of the land breeze for their return. Scores of small white sails flashed in the rays of the western sun. They were a sight to remember. And here was Papa Jean, standing with his bare feet wide apart, his pants rolled up to the knees, his ragged and sleeveless shirt hanging open in front, and a string of sparkling glasslike fish in his hand. The fish were hung on a switch that went through their gills.

"What will we do with so many fish?" Fifina asked as she looked first at Papa Jean's string and then at the crabs.

"Some of them we'll eat," he said. "Some of them we'll sell in the market. And if we have any left I'll carry them to your Uncle Jacques, who lives at the other end of rue Bord de la Mer. We have not gone to visit him since we moved to town, and I'd like to take him a present."

"May I go too?" Popo asked hurriedly.

"Not to-night," Papa Jean said. "It will be too late when I get back. But maybe another time. . . . Hello!" He suddenly noticed the string of crabs at his feet on the ground. "Well, just look at these! I see I have two bright children. Won't Mamma Anna be happy to see these! Some day I'm going to give you a treat. Maybe next Sunday I'll take you for a walk to the light-house."

Popo put his shoulders back. He felt as big as a man. And the promise of a trip to the light-house made him forget for the moment that he had just been denied the trip to Uncle Jacques's home. He followed Papa Jean and Fifina up the slope, dragging his string of crabs just as proudly as Papa Jean carried his sparkling fish.

V

A TRIP TO THE COUNTRY

A FEW weeks passed and Popo began to feel that the family belonged in the town. He almost forgot that just a short time ago they had been strangers in Cape Haiti. But now he heard talk that took his mind back to the country. Mamma Anna was homesick for her relatives in the hills. She was anxious to see her mother, Grandma Tercilia, again. She wanted to be there in the house that always seemed like her real home; it was, in fact, the house in which Mamma Anna had been born. She was lonesome too for Aunt Marie and her large children. Mamma Anna had never before been away from her relatives, and it was natural that she should be homesick.

"You may go to the country then Saturday with the children," Papa Jean told her. "That will give you two days to visit. I will come for you late Sunday afternoon."

With that permission everybody became happy. Popo danced in the middle of the floor. Fifina clapped her hands. And Mamma Anna hid her face to keep from showing that her joy had made her cry.

Saturday! Oh, happy day!

Why don't you hurry and come, Saturday? Can't you see little black Popo sitting in the hut door waiting? Can't you see Mamma Anna standing silently with Pensia in her arms, waiting? Why don't you hurry, Saturday?

But Saturday was hurrying. Saturday came. Popo put on his little white shirt and felt very dressed-up. Fifina put on her cleanest dress and Mamma Anna gave Baby Pensia a good bath. And once again the little band started on a journey. This time the trip was unlike the last one they had made, in several ways. First they had left the burros in the back of the yard, for there were no large bundles to be carried. And secondly, Papa Jean was not with them. He had gone out in the boat as usual, for he could not afford to miss a day's work.

At the edge of the town they paused to notice the slaughterhouse. This was a concrete platform under an iron roof. Here every day sheep and cattle were killed so that the people of the town might have meat.

One might think that in a town as small as Cape Haiti it would not be necessary to kill animals every day. But in Haiti ice is very hard to get. There are no ice boxes or big refrigerators; and since meat will not keep in so warm a climate, the animals must be killed every day. For that reason the slaughterhouse is always busy.

Mamma Anna paused a few moments to let

her curious children get an idea of what was going on. There were a dozen bulls and sheep tied near by, and men were sharpening knives; but they did not wish to see the poor animals put to death, and so they soon turned their backs on the scene.

A little farther down the road they passed the only important factory of the town, where pineapples are canned and prepared for shipment to the United States. Here Popo saw many black men working. They did not move about leisurely, like other workers, and Popo thought he wouldn't enjoy working so hurriedly.

"Where shall we sleep to-night?" Fifina asked her mother.

"Perhaps with Aunt Marie," Mamma Anna said. "Grandma Tercilia hasn't so much space now. But at Aunt Marie's we may be able to spread mats on the floor for you children. She will have enough space."

"I'd rather sleep at Grandma Tercilia's," Popo said.

"I wouldn't," Fifina said. "Grandma Tercilia has that pig that runs in and out of the house all the time. He is a terrible nuisance when you have to sleep on the floor."

"The pig sleeps, too," Popo argued. "He would not be able to trouble us when he's sleeping."

"Well, anyway we'll sleep at Aunt Marie's this time," Mamma Anna said finally.

For the rest of the journey they walked in silence. They passed down a long road shaded

by mango trees and banana plants, tall prickly cactus and high palms. And in the afternoon, tired and dusty, they came to the little by-lane through the coffee bushes that led up the hill to the huts of their relatives.

First they went to Grandma Tercilia's. Popo could see from a distance the tiny thatched house that he knew so well. He could see the great mango tree weighed down with fine fruit, beautiful ripe mangoes that were now orange and greenish—a charming sight. Certainly it was pleasant to be back, Popo thought. He could hardly wait till he reached the top of the hill. His mouth began to water when he remembered how sweet the mangoes from his grandmother's tree were. Surely not another tree in Haiti bore finer ones. He and Fifina broke away from Mamma Anna and scampered up the path.

Grandma Tercilia, an old wrinkled black woman with a pipe in her mouth and skin that

was parched like an autumn leaf, met them at the door.

"*Ah, mi cher, ti monde!*" she exclaimed. It was her most endearing expression, spoken in Creole French. It meant, "My dear little ones."

In a few minutes children came from every direction, from around the hut, from the banana trees, from the thickets, all of Grandma Tercilia's children and grandchildren who still lived at the family home came out to meet the visitors from town.

Grandma Tercilia and Mamma Anna began talking so fast and excitedly that Popo could hardly tell what they were saying. Then they fell upon each other's necks and embraced as if they had been separated for many years instead of a few short weeks.

A few minutes later the whole family came out in front of the thatched hut and sat on the ground. Popo and two or three other children climbed the mango tree and shook the branches. A rain of fruit fell. The children gathered them from the ground and passed them to the old folks. Then everybody began to eat.

The mango is a juicy and rather sticky fruit with a stone to which its meat clings. The skin is rather tough and is peeled back as the fruit is eaten. The meat of the mango is bright yellow, and to Popo it tasted good enough to repay all the trouble of eating it.

As the folks ate, they tossed the skins and seeds to a greedy little pig that ran from one to another, grunting and never seeming to be satisfied with what he got. Some fighting cocks belonging to big Cousin André pecked about, enjoying the fruit skins, too.

By and by, every one was through eating, and Fifina remembered her job. She went to the spring a short distance from the house, took the gourd that was left there always for convenience and brought it full of water for the old people to wash their sticky hands.

Popo followed her on the path and noticed the great bright-winged butterflies fluttering above the bushes along the way. His heart was light and happy, but he was beginning to feel sleepy. He had got up very early in the morning, and with Mamma Anna and Fifina had made a long and tiresome journey.

When he came back to the hut, he stretched out on the ground in front of the door and went to sleep. When he awakened, the sun was out of sight. Things were quiet around the house, and

Mamma Anna suggested that it was time to start for Aunt Marie's place if they wanted to get settled before it was pitch-dark.

As they walked beneath the banana plants, Popo amused himself throwing stones at the dark lizards scurrying among the leaves and grasses on the ground.

VI

DRUMS AT NIGHT

THIS time Popo could not go to sleep. Baby
Pensia, Fifina, Mamma Anna, and his aunt
Marie were all dreaming in the little hut. But
Popo lay awake looking through the cracks in
the door at the moonlight streaming down the
mountain slope outside. He was listening to the
drums in the valley below.

It was at least a half-hour's walk to the place
near the main road where at night the drummers
played and people danced the Congo. Popo had
never been down there, but big Cousin André
had gone there now, he was sure, because Mamma Anna and he had met him going down the
path with some neighbor boys just after sunset.
Popo wished he were a big boy so that he could go
to dances, too. Indeed, as he lay on his little straw
mat, listening to the drums, he felt sure he could
find his way down the hill and up the main road
to the place where the drums were. In the quiet
night, they sounded quite near, booming deep
and quick, in a lively sound that made your feet
want to keep time. Through the wide door

The drums sounded louder and louder.

cracks he could see that the whole valley was a pool of moonlight. And the drums and the dancers were just at the foot of the hill.

Popo got up and went out. Two men were passing down the mountain path in front of his aunt's hut, laughing and talking. He knew they were going to the dance, and since André was there, Popo decided at once to go, also. What fun it would be, watching the drummers play! Maybe they would let him hit the big drum once. Anyhow, he would see. So down the mountain trail he started, not far behind the men.

The banana leaves were like long green fans in the moonlight. The wind made them sway languidly. Sometimes, on the way down, a palm tree would shoot up very tall, with its head against the stars. Once he crossed a gurgling brook that seemed to be in a great hurry to get somewhere. Popo knelt in the very middle and took a drink. How cool the water was on his knees and chin! He liked water so much that he would have lain down in it, if he had had time. But he had to hurry on to the dance.

As he came into the valley, the drums sounded louder and louder. The path was wider now, and the moon was bright overhead. The sky was full of stars. Several grown boys and girls were coming behind him, and before he knew it, they were all on the main road, where there were huts and lights and many other people going to the dance. Nobody paid any attention to the little black boy walking along by himself. Popo didn't mind that, for children in Haiti are used to walking alone in the dark. He knew the night was as kind as the day. And near at hand were dancers, drummers, and drums!

Ahead, was the flicker of many kerosene flares by the roadside. As Popo drew near, he could see under them the little stands of women with piles of candy to sell, white buns, hot fish and fried yams, and little pots of black coffee over charcoal fires at their feet. He wished he had a penny to buy a bun.

Popo wended his way into the grounds, where, under a thatched roof open on all four sides, many people were dancing, and at one corner, three black men were playing on tall drums of different sizes, swaying back and forth, their hands moving at a rapid rate, beating out the deep vibrant music.

The drums, as Popo knew, were made from the hollowed-out trunks of trees, one end covered with the dried skin of a cow or a goat, sometimes with the hair of the animal still left on the head of the drum. The tallest drum was about four feet high, just about the size of Popo. A huge young man held it between his knees, slanting the drum away from him, as he played it with a stick in one hand, while the fingers of his other hand brought forth a swift series of continuously happy booming sounds. The two smaller drums were played by men who used their fingers alone to make the music. Beside one of the drummers a boy squatted with a pair of sticks in his hand which he kept beating in time against the long wooden sides of the big drum. The three drummers, and the boy with the sticks who beat on wood, were all rocking gayly back and forth as they played, pleased to be making such grand music for the dancing people under the thatched roof.

Men and women were there dancing, facing one another but not touching, their feet scooting across the floor, heads up, hands out, and faces smiling under the oil lanterns that made the dim light. High over the thatched roof and the banana trees and the palms, the sky was bright with stars and moon.

Popo stood very close to the drummers, thrilled by so much noise and so many people. It was even livelier than the market at the Cape on a Saturday afternoon, for here everybody was moving in time to the music of the drums, laughing and dancing. Through the crowd Popo saw André dancing all by himself, twirling round and round, and crossing his feet in quick rooster-like movements. Popo tried it, too, but just as he began to get the steps right, the drummers stopped playing and the dancing ceased. Then the little boy ran through the crowd, calling, "Hello, André!"

His cousin seemed surprised. "I thought you were asleep," said André.

"I was, almost," Popo replied, "but I heard the drums."

"Well," said André, "you had better go back home before Anna wakes up, and finds you gone. She will be worried about you."

"I'll go when you go," Popo said.

"That will be soon," André replied. And before Popo had a chance to ask to beat the big drum, his cousin took him by the hand, bought him a cupful of peanuts from one of the roadside stands, and led him back up the mountain path.

"Little boys should be asleep," said André.

"And you, too," Popo replied.

André laughed.

All the way up the hill under the banana trees, and across the gurgling brook, they could hear the drums beating happily in the valley below.

VII

PLAY

ONE afternoon when Popo and Fifina were on the beach waiting for Papa Jean to return in his little boat, they noticed that the bright blue sky was full of kites. The kites were flying high and smoothly, like a flock of sea gulls on a gentle wind. Some of them were square, and others were triangular, and still others had long tails such as were never seen on real birds.

"Oh, Fifina!" Popo cried. "Did you ever see anything like them?"

"Never," Fifina said. "Look at that big one like a box. What a strange kind of kite!"

"And the little one without a tail—isn't it a beauty?"

"But where are the strings? And who is flying them?"

They searched the beach with their eyes, looking up and down as far as they could see; but they could see no one. The kites seemed to have no anchoring strings. They seemed indeed to be as free as the tropical birds of Haiti which they resembled so closely.

A large white cloud hung above the horizon

like a puff of smoke. It was a fine sight. Water lapped the clean sand of the beach and ran to where Popo and Fifina stood—a long green tongue of water licking the sand and wetting the feet of the children.

Suddenly Popo shouted:

"There they are, Fifina. See, on that big rock away down the beach. See the boys holding the strings. They are the ones who are holding the kites."

"Yes, yes. I see."

"I'd love to have a kite, Fifina."

"Maybe Papa Jean will make you one."

"Do you really think so, Fifina?"

"I do," she said. "I think he will gladly make you one if you help around the house and give Mamma Anna no trouble."

Popo bowed his head and began to think. He was very anxious to be helpful and to make a good impression on Mamma Anna. He knew that if Mamma Anna approved of his having a kite Papa Jean would be much more likely to grant the request. So he wrinkled his brow and wondered how he could help his mother.

"Let's go home, Fifina," he said finally. "I want to see what I can do for Mamma Anna."

"We might go to the fountain and help the milk women with their cans. They may give us a penny, and if they do we can take that home with us. That would be a good thing to do. Mamma Anna needs money."

"That's fine. You are almost as smart as a grown person, Fifina."

She shook her finger before her face play-fully.

"You are teasing me, *petit monde*," she said affectionately. "I don't always think of smart things."

They ran up the slope to the rue Bord de la

Mer. Then they turned toward the public fountain and began walking in the middle of the street. Popo kicked his feet in the soft dust and scampered off ahead, but Fifina was like a little lady. She walked sedately, swinging her arms.

Sure enough there were two women at the fountain with their milk burros. These were

the animals that carried their cans. On each
burro's back there was a grass mat like a blanket.
And on either side of the mat there were large
pockets into which the milk cans fitted.

The women had evidently finished their day's
rounds and had come to the fountain to rinse
their milk cans with clear water before carrying
them home. They followed their tired little
beasts up to the water spout and began removing
the large empty cans from the pockets of the
mats.

Popo stepped up very respectfully.

"Help you wash the milk cans, ma'am?"

The tall strong woman looked down at the
tiny black youngster at her side. She smiled a
little.

"Why, son, you are not as big as one of these
cans."

"But I think I could wash them for you," he
insisted. "I know how to wash things. Mamma
Anna always lets Fifina and me do her pots and
pans."

"I am tired of walking," the woman mused.
"I have been on my feet since early morning
and I could enjoy a little rest all right." She
looked down at Popo again. Then suddenly she
made up her mind. "Well, go ahead and try it.
If you wash them well, I'll give you a penny."

Popo did not need a second invitation.
Quicker than you could say it he had turned the
huge can over on its side and was inside it, head
and shoulders, with a handful of sand scrubbing
its sides. Popo certainly knew how to scrub

things at the fountain. His little naked body wriggled like the hind legs of a frog as he twitched and worked with his hands. Popo was a sight.

In the meantime Fifina had gone to work on the cans of the other woman. She did not tip

the can over on its side as did Popo. Neither did she crawl into it with her head and shoulders. Instead she reached down with her arms, as a grown woman would have done. By standing upon the little concrete step in front of the fountain she was able to reach nearly to the bottom.

While the children worked, the two women sat across the street beneath a tree. They dug the toes of their bare feet into the ground, kicked up dust, and stretched their tired legs. A moment or two later they were both smoking black cigars and chatting happily together. Popo caught a glimpse of the pair when he came out of his can for a fresh handful of sandy dirt. He saw their heads wag as they laughed, saw their pearly teeth flash.

Perhaps they were talking about the new hats they hoped to buy when they had saved enough money from the milk business. Or perhaps, he thought, they were laughing at the fat man across the street. He had just tipped his hat at them. And such a hat it was! A high silk topper such as old-fashioned coachmen used to wear. And the funny thing was that the rest of the fat old fellow's clothes were pitifully ragged. The milk women hid their faces with their hands when he stared at them.

When the four cans were clean and shiny inside and out, the women gave each of the children a penny as they had promised, and Popo and Fifina set out joyfully for their home. The sun was slipping down the sky fast now. The tide was coming in, and little white breakers were splashing on the rocks along the water front. Out on the clear green water there was a line of silver sails. All the dozens of small fishing boats were quietly returning with their day's harvest of fish. The fishermen of Cape Haiti were bringing in the harvest of the sea.

Their sails flashed in the sunlight like white wings.

"There's a pretty sight," Fifina said, pointing out across the water.

"Yes," Popo agreed, "but where are the kites, the pretty kites we saw before we went to the fountain?"

"I guess they have been taken down. And I suppose the boys have gone home for dinner."

"Already?"

"It's late," she said. "The sun will be down before you know it."

"Well, if I had a kite I would not take it down so soon on a fine afternoon like this. I'd leave it up as long as I could see."

"You had better not talk like that to Papa Jean or Mamma Anna. They might not be so anxious to make you a kite if it's going to keep you out when you should be home having your supper."

Popo thought for a minute. That was an idea. A little boy could not always tell all the things that came into his mind. Old people had so many strange ideas. But Fifina understood them all. She was certainly a great help.

"Oh, he said finally, "I see. I won't tell them how long I'll leave it up."

VIII

THE NEW KITE

PAPA JEAN rubbed his chin with the back of his hand. His forehead wrinkled. He was thinking about Popo's kite. What chance would a busy fishing man like Papa Jean have to bother with a kite? But, on the other hand, how could he refuse? There were Popo, Fifina, Mamma Anna, and Pensia all looking him in the face and waiting for him to say yes.

"They have been fine children to-day," Mamma Anna said. "You should make a kite to please them, Papa Jean."

Papa Jean threw up his hands in despair.

"All right, all right," he said finally. "Just leave me alone, and I'll make you a kite. If I don't, I know I'll never have another minute's peace. When you wake up to-morrow, you will find your kite finished and ready to fly."

Popo sprang up from the dirt floor, kicked his feet in the air.

"Oh, Papa Jean, Papa Jean!"

Fifina clapped her hands and danced.

"Oh, Papa Jean, Papa Jean!"

Mamma Anna smiled. Pensia cooed.

Oh, Papa Jean, Papa Jean! How happy you have made Popo and Fifina! A real kite! Just think of it. A beautiful red or yellow or green kite, like a bright bird of Haiti. A lovely delicate kite to soar in the clouds—like a wish or a dream. Oh, Papa Jean, how do you expect Popo and Fifina to sleep to-night.

The next morning the children were up at daybreak. And sure enough, there was the big handsome kite Papa Jean had promised. It was red with trimmings of yellow and green paper. Beside it was a large ball of strong cord. Popo examined everything carefully. This kite *was* a dream.

"It's all ready to fly, Fifina."

"Yes, Popo," she said. "And we couldn't have wished for a better one."

They carried the treasure out between them. They walked slowly, carefully down the slope to the beach. When finally they stood upon the clean white sand, they paused to make sure that all was well overhead.

No, not quite ready. Popo shook his head. There was a tree limb a short distance away, and he did not care to take a chance. He had seen many a fine kite hung in the branches of a tree. And he did not care to lose his in the same way.

"Let's move a little farther down the beach," he suggested.

"All right."

Fifina helped him bear the kite beyond danger.

"Now," he said. "You hold it up while I let out the string. Hold it lightly, and when I tell you to, let it loose. See?"

"I understand," she said. "Hurry up and unwind your cord."

Popo let out about fifty yards of the string. Then he looked behind himself to make sure there was a clear running space. Everything was clear.

"Ready," he cried. "Let her go."

At that instant Fifina released the kite. And at the same time Popo began to run.

Almost immediately the lovely bright thing began to climb up into the air, a big scarlet star rising from the seashore. How wonderful! Up, up, up, steadily it climbed. Popo began letting out more cord. He had a large ball, and he could afford to let it out generously.

Fifina ran to his side, but she did not speak. There was nothing to say. What could anybody say who saw a great red star rise from the white

beach in the daytime? It was just beautiful to
look at, and the less you tried to talk about it
the better you could enjoy seeing it. But Fifina
knew that Popo's kite was not a real star. It just
looked like a star and made her think of one.

Across the harbor the tiny sailing boats could
again be seen. Another day of toil had begun
for them, another day on the tossing water. In
one of them stood Papa Jean. He was leaning
with one arm against the little mast and the
other arm waving proudly in the air. As he stood
there, the wind went through his ragged shirt
flapping it like a bullet-torn flag. His naked
arms and shoulders flashed like metal in the
early sun.

Popo knew that Papa Jean was as proud as
anybody of the kite. And why shouldn't he be?
Hadn't he made it?

Soon all the little boats were out of the harbor. The bay became still and the round copper
sun mounted the sky steadily.

"How is the kite pulling?" Fifina asked.

"Fine," Popo said. "The wind is still good up there, I guess."

"H'm. I guess it is. Does she hum?"

Popo put the cord to his ear. His face brightened.

"Hum?" he cried. "Hum? Why, this kite sings."

"Let me hear."

Popo held the string to his sister's ear. "How's that for singing?"

"It's good. Let me hold the string awhile."

The two youngsters walked down the beach to a place where some stones jutted out into the water. There they sat down to rest, and held the string of the kite in turn, feeling the steady firm pull and vibration, listening to the soft purring hum of the string. It was great fun.

Presently Popo looked up and saw that there was another kite in the air. When had this second kite risen? Who was flying it? And what did the stranger mean anyhow by cutting in on Popo and Fifina? There were other places to fly kites. But the strange kite, a dull brown thing, rode the wind just as gayly as did Popo's. It ducked and darted about so wildly that Popo feared it would become entangled with his own string. It reminded him of a hawk swooping over a smaller bird.

A moment later something happened. The big brown hawk-kite got across Popo's string, and began ducking and darting more than ever. About that time Popo caught sight of the boy who held the string. It was clear that the fellow was very proud of his rude misbehaving kite. He was jerking the cord and plainly trying to saw across Popo's string and cut it loose. It was an old game, and Popo knew something about it. Mischievous boys often cut the strings of innocent youngsters, causing them to lose their kites in the sea. But Popo had confidence in the kite Papa Jean had made him. He believed his big red star-kite was a match for any hawk. He started jerking vigorously on his cord. Back and forth, back and forth. Then suddenly something happened. Popo's heart stood still.

A cord snapped. Popo could feel it go. But it was not his cord. His kite pulled and sang as steadily as ever; but the other one, the hawk, was falling to the earth like an evil bird with a broken wing. Down, down, down it sank. A moment or two later it dropped into the ocean. Popo's big red star climbed the sky proudly, a true conqueror.

IX

A JOB FOR POPO

A FEW nights later Papa Jean and Mamma Anna sat in their doorway calmly smoking pipes. The children were still on the beach flying their kite. For three or four days they had done almost nothing but fly that kite, and Mamma Anna had begun to wish that she might have them at home more. Often she had little things to do, but when she called, nobody answered. Realizing that they were not in the yard she would go to the door and look into the sky. When she spied the big red kite, she would shake her head in despair. That meant that they were half a mile down the beach, and probably would not come home for hours. It meant that Mamma Anna would have to go to the fountain herself for water. In the evening Papa Jean would look about for some one to go with him to the market and help him sell his fish, but he would see no one. His children were not to be found. But in the sky he could always see the red star-kite he had made them.

As the parents sat in their doorway smoking, the voices of the youngsters suddenly rang out above the waves, and a moment later they came up the slope, tired, sleepy, exhausted, but happy. They never seemed to get enough of flying that kite. Once again they had kept it up till the night was dark.

Coming toward the house, Popo heard the voices of his parents. They were talking very softly, very calmly, but in the still night air their words were perfectly distinct.

"All play and no work makes Jack a dull boy," Papa Jean was saying.

"H'm," Mamma Anna agreed. "That's an old foreign saying, but it is very true."

Popo knew what they meant. He realized that for several days he had done nothing but play, eat, and sleep. He had been wonderfully happy. He had forgotten all about work. These words made him remember.

"Well," Papa Jean said, changing the subject quickly, "did you fly the kite again to-day?"

"Did we?" Popo swelled with pride. "Tell him, Fifina."

"We let out all the string to-day," she said. "We had it so high it looked like a tiny speck."

"Ah, that's good."

"H'm," Mamma Anna said. "That's the way to fly a kite."

"And how it pulled!" Popo exclaimed.

"H'm, and how it sang!" Fifina cried.

"That's very good," Papa Jean said.

"Very good," said Mamma Anna.

"But I've been thinking." Papa Jean rubbed his chin.

"We've been thinking," said Mamma Anna.

"I've been thinking that maybe you have flown the kite enough for a while. Maybe a little work would be a change."

"H'm," said Mamma Anna. "A little work would be a change."

For a moment no one spoke. Popo sat on the ground in the moonlight, his kite across his

knees. Fifina sat beside him holding the ball of cord. Down at the bottom of the slope the waves were beating against the rocks.

"What can we do?" Fifina asked humbly. "We like to work and help."

Popo said nothing.

"That's a good girl," Papa Jean said. "What about you, Popo?"

"Oh, yes," he said quickly, "I like to help."

"Well, to-morrow we'll leave the kite on the shelf."

Popo brushed a tear away.

"All right, Papa Jean," he said.

"Fifina will help Mamma Anna in the house. Popo is growing fast. I think it is time for him to have a real job."

Popo felt his heart leap. He had not guessed that he would be going to work like that. He had thought that Papa Jean and Mamma Anna only wanted him to run errands around the house.

"Will you take me with you in the boat?" he cried.

"No, son, not yet. I am going to take you to your Uncle Jacques's cabinet shop. You can stay there and learn the trade. It will be better than fishing. That is, it will give you two trades. You can learn the fishing later, any time. I can teach you that."

Mamma Anna kept nodding and humming as she agreed with the things Papa Jean said. H'm, h'm, h'm-m. It was almost like some one singing.

"You will have to be up early in the morning," Papa Jean continued. "I shall have to take you to the wood shop before I go out in the boat. So you had better be getting to bed."

"All right," Popo said.

He and Fifina went into the house. A moment later Mamma Anna came in and lit a flare that filled the room with a soft yellow light. And almost before you could say the word, they were tucked in their sleeping places in the corner.

The next morning Popo and Papa Jean walked nearly a mile down the rue Bord de la Mer. At the end of their walk they stood in the entrance of a small woodworking shop that smelled pleasantly of fresh-cut wood. There were several workbenches in the room and a number of large saws and tools for cabinet-making. Against the rear wall was piled the rough sweet-smelling wood.

Work had not yet begun in the shop, and only an old frail man was there. This was not Popo's Uncle Jacques, but Uncle Jacques's helper, old man Durand. The wrinkled old fellow was very ragged. He was stiffened with age and rheumatism and got about almost as slowly as if he had been actually lame. With his toothless gums he chewed on the stem of a cob pipe, and at the moment that Popo and Papa Jean came to the door he was busy moving things about, getting ready for work. He looked around when he heard the visitors enter the door. He was very friendly, and when he smiled his old face split up like a pie into which some one had punched his fist.

"Is my brother here yet?" Papa Jean asked.

"He was here," old man Durand said. "He has gone back home a minute to drink a cup of coffee."

"I'm in a hurry," Papa Jean said. "I suppose I'd better go to his house and call him."

"Not at all," old man Durand said. "Just stay right here. I will get him for you in a minute."

The old man went out into the street and started down the block. The sun was not yet up, but a pearly gray light hung along the horizon. Smoke was coming from the windows and little chimneys of some of the houses. The air was full of the fine odor of fresh coffee.

Soon old man Durand came hobbling back, followed by another man, a tall strong fellow who was much like Papa Jean in the face. This was Uncle Jacques. And trotting along behind him was a small boy. This was Popo's cousin, Marcel.

The two brothers greeted each other. Uncle Jacques was older than Papa Jean, and their resemblance was more nearly that of father and son than that of brothers. Marcel and Popo stood behind their respective parents. Though they were cousins and had heard of each other, they were timid at their first meeting.

"I've brought Popo to put him to work," Papa Jean said.

"Fine," said Uncle Jacques. "My Marcel has just started to work, too. They can learn together. I had two other boys, but I let one go last week to make a place for Marcel. I will let the other one go to-day. One has to favor his kinsfolk."

"Thank you, brother. I'm leaving Popo in your care. See that he works hard and learns the trade. He will be yours till he has learned enough to earn money at woodworking."

"Fine!" Uncle Jacques said, putting his hand on Popo's shoulder.

"Now I must be gone. It is time the boat was getting under sail."

"Don't worry, Jean. Popo will be at home here."

Uncle Jacques threw his hand into the air as Papa Jean disappeared down the street. The sun peeped over the horizon.

X

OLD MAN DURAND

POPO stood beside his workbench smoothing a board that was to become the top piece of a little table. Uncle Jacques had shown him how to put his sheet of sandpaper around a block of wood and then use the block as if it were a plane. The thing worked like magic. It slipped easily over the surface of the rough board, and wherever it touched it left the surface smoother. Popo was happy. He was helping to make a table— little Popo helping to make a table for the front room of some well-to-do family of the town. And what he was doing to the table was important too. Why, what would a table look like without a fine smooth top?

While he was working, and thinking these thoughts, he did not pay much attention to the youngster who was working at the bench directly in front of him. But a moment later, when Popo stopped to put another sheet of sandpaper on his block, he saw that the other boy had paused a moment also. Marcel was looking at Popo with a pleasant smile.

"Well, how do you like it, cousin?"

"Oh, fine," Popo said.

"That little table you're working on isn't much. Anybody can make a table. But it's fun when you learn to make really pretty things."

Popo didn't like what Marcel had said. It sounded as if Marcel had no respect for the simple, easy things that Popo could work on.

"Maybe I can only work on tables now," Popo said sharply, "but I'll learn as fast as you. I'll make things that are as pretty as anybody can make when I have had a chance to learn."

Marcel laughed out heartily.

"Sure you will, cousin. I mean it's not much fun when you are just beginning, when you have to just sandpaper boards and do trifles like that."

"It sounded as if you were making fun of me."

"Never, cousin. Come here a minute—I'll show you something."

Popo went around to Marcel's workbench. Before him lay a beautiful serving tray carved from a single piece of wood. Marcel held it up so that Popo could look at it to advantage.

"It's a beauty," Popo exclaimed.

"Oh, it's just plain," Marcel said modestly. "You see, I haven't been here so long myself. This is the first pretty thing I have tried to make."

"And shall I make things like this?"

"Surely, cousin. You'll make a tray like this in a week or two. Trays are easy."

"But all that little fancy business around the edge and by the handles—isn't that terribly hard?"

"That is about the hardest part, but Papa Jacques will help you do that on the first one. He helped me with this one."

Popo fixed his eyes on the careful designs carved by hand in the wood. There were flowers, leaves, and stems. The handles were twisted like the coils of a vine.

"Does Uncle Jacques have another tray, a finished one, that he goes by when he makes a new one—a pattern?"

"No," Marcel said. "The trays are all made alike, but each design is different. So the old patterns would not be much good. If a lady wants to buy a tray, she does not want it to look exactly like the one her neighbor has. So each one has to have its own design."

Popo went back to his own bench and worked on his board.

Sunshine flooded the little workshop. Uncle Jacques looked very tall and dignified working near the door. His forehead was wrinkled as he leaned over the strips of woods that he was fitting together. And as he worked, Popo noticed, he whistled or hummed a tune.

Old man Durand kept hobbling around, moving things and smoking his pipe. He was making a stool. But old man Durand was forgetful like many other old people, and he could never remember where he had laid his tools. This kept him constantly busy. So during the course of a day, old man Durand did not get as much done as Uncle Jacques—although he was really just as good a worker.

One thing was not clear in Popo's mind. How could a person make designs on a tray without a pattern? And how could he get new designs for each tray? It was hard to figure out. He wanted to make all the fine pieces of furniture that Uncle Jacques and old man Durand made, but he was afraid it would take him a long time even to make an attractive tray. Later in the afternoon he went over to Marcel's bench again.

"Marcel," he said quietly, "tell me this: How can anybody make a design without a pattern, and how can he make a new design every day for every new tray he works on?"

"That's a hard question, cousin. I've wondered about it myself. Let's ask Papa Jacques."

They went over to Uncle Jacques's bench.

"Uncle Jacques, tell us how you make designs without a pattern, how you make a new design each day, a new one for each new tray."

"That's a hard question, boys, very hard. Ask old man Durand."

Old man Durand was pounding on his chisel with a wooden mallet. When the boys stood in front of him, he looked up and wiped the perspiration from his forehead. His face split up in a great smile. He took his pipe from his mouth and held it in his left hand.

"Old man Durand, tell us this: How do you make a design without a pattern, and how do you make a different design every day?"

"That's a hard question, boys, very, very hard."

Old man Durand.

"Yes, old man Durand, but you must tell us. We want to know."

"Well, boys, it's like this: you have to put yourself into the design."

"Ah, you're teasing us, old man Durand. That's a riddle. How can a boy put himself into a design?"

"Ah! It's a riddle indeed, but I'm not teasing you. If I walk down by the beach on my way to the shop in the morning and see the tiny boats putting out to sea, that makes a picture in my mind. If I see a hungry beggar, that leaves a picture too. Some pictures make me glad to be living. Some make me weep inside. Some make my heart sad. And when I'm glad to be living, trees and birds and leaves look one bright color to me. When I weep inside, they look different. Well, I don't think about this when I sit down to make my design, I just sing or whistle a tune and carve away with my knife or chisel. But what I am inside makes the design. The design is a picture of the way I feel. It sounds strange, but it is just like that. The design is me. I put my sad feeling and my glad feeling into the design. It's just like making a song."

"It's wonderful," Popo said.

"It sounds just like old folks, but I like it," Marcel said.

"And when people look at your design," old man Durand went on, "when people see the picture, they will just see trees and boats and flowers and animals and such things, but they will feel as you felt when you made the design. That's the

— 73 —

fine part. That is really the only way that people can ever know how other people feel."

"There is nothing in the world like making designs!"

"Nothing is finer," old man Durand said.

"That's true," Uncle Jacques agreed. "Old man Durand has told you a beautiful thing and a very true one. I hope you understand him."

"I am sure I do," Popo said.

"I too," said Marcel.

"To-morrow, Popo, you shall start on a tray of your own," Uncle Jacques smiled.

"To-morrow!"

"Yes, to-morrow. You are a bright boy, and I think you can begin one right away."

Popo was so excited and happy that he could not speak again. To-morrow he would make a tray, a beautiful tray with a design on it. And there would be nothing but happiness in that design.

XI

POPO'S TRAY

WITH his hands stuck proudly into the pockets of his new breeches, Popo walked to work like a man. He felt like a man too. The morning air was fine. People were in the streets. Some were going to the fountains; others were starting out for their jobs. Popo was glad to be among them. He too was going to a job. He was going to a job he loved.

On the sidewalk he saw a group of men sitting beside some small tables that had been placed outside in the sunshine by the proprietors of a café. The men were playing an odd game of cards, in which the losing pair had to wear clothes pins on their noses. When they began winning, their opponents had to wear the clothes pins. Popo thought it very odd for old men to wear clothes pins on their noses while they played seriously at cards. But he could not stop long to watch them. He was in a hurry.

Old man Durand was already in the shop. The old fellow had swung the big door open and was getting things in order.

"You are early, Popo," he greeted the boy.

"Yes," Popo said. "I came early to work on my tray—the one Uncle Jacques started me on —I want to finish it to-day."

"I am sure you will," old man Durand said. "And I think it will be a mighty fine one too for the first you ever made."

"I hope so," Popo said. He tried to talk so as not to show how eager he was to make a success of his first tray.

Old man Durand lighted his pipe and set to work at his own bench with a mallet and chisel. Popo began sanding his tray. A little later Uncle Jacques and Marcel came in. Popo looked up and greeted them pleasantly, but he did not stop working. He was making a picture of a sailboat in the bottom of his tray and all around it he had drawn curves to represent the waves of the ocean.

The hours passed. Still Popo worked. Soon it

was lunch time. Popo took only a few minutes off. Then he hurried back to his workbench. He would have to hurry to finish his tray by evening, and he needed every minute for work.

That afternoon, while they worked in the shop, Uncle Jacques told a few stories to amuse the boys. Popo did not stop work for a minute, but he heard what Uncle Jacques said. That is, he heard it in snatches; and he remembered the pictures Uncle Jacques drew as he remembered dreams. Uncle Jacques could tell stories well.

It seemed that the grandfather of Uncle Jacques and Papa Jean had lived to be more than a hundred years old. Before his death he had told his grandsons some things he remembered. He had told them how the great stone fort called the Citadel of King Christophe came to be on the high, far-away mountain overlooking the town of Cape Haiti. For old Grandfather Emile, as a boy, had seen it built.

Away back in those old days many heroic things were done. Grandfather Emile had seen the black workingmen drag large bronze cannon through the streets to the foot of the mountain. Later he had seen them drawn up the steep sides of the hill by an army of half-naked people who tugged and pulled like animals. Grandfather Emile had told about men who left their homes to work on the Citadel and remained away for ten or twelve years at a time without returning to their families for a single holiday. Once, he had seen the great black king who was responsible for this huge structure. He had seen King

Christophe pass through the town on a white horse, surrounded by bodyguards in flashing uniforms.

That was wonderful, Popo thought. It was a pretty story. But somehow it made Popo feel sad. He could see the ruins of the Citadel from the door of the workshop, and it made him sad to think how hard people had worked to build it. He could not forget those poor men who dragged the heavy cannon up the mountain, or those others who went away from home and could not get back for ten or twelve years.

But as Uncle Jacques went on with his story, Popo began to understand. The Haitians had once been slaves to the French. They had freed themselves, fighting. Then they had built that fort, the Citadel, as a protection, so that the French might not come and make them slaves again. And that was why the men worked so hard, and stayed away from home so long.

Popo began thinking about his tray again. He must not waste any time. He must hurry if he wanted to finish it by evening. He thought of what old man Durand had said about a design being a picture of how you feel inside. And Popo wondered if his tray would show that he felt sad as a result of Uncle Jacques's story about the men of old. He hoped not, for it was his plan to have nothing but happiness in his first tray. He wanted everybody to know how glad he was to make something with his own hands.

Soon twilight began to creep into the workshop. Popo put the last touches on his tray of wood and held it up with a cry of joy.

"It's finished, Uncle Jacques. It's finished, Marcel. It's finished, old man Durand. Just look. All finished."

"Ah! That's fine," said Uncle Jacques, noticing the pains Popo had taken.

"Ah," said Marcel, looking at the sailboat in the center, "that's fine!"

"Fine!" said old man Durand, gravely.

"I'm happy," Popo said. He felt as if he could cry, but he was too big for that. Besides, what was there to cry about?

"Since that is your first," Uncle Jacques said, "you may carry it home and give it to Mamma Anna. Tell her to take good care of it. Tell her to remember that it was your first piece of work in my shop. When you are older you will make many beautiful things, many much finer than this tray; but there will never be another first one. For that reason the first one is precious."

"Thank you, Uncle Jacques," Popo said. "I'll tell Mamma Anna what you said."

"And remember the riddle about putting yourself into your designs," old man Durand said with a grin.

"I'll try," Popo answered.

XII

MAKING PLANS

ONE day, after he had been in the shop several weeks, Popo said to Marcel:

"Have you ever gone to the lighthouse, cousin?"

Marcel looked up from his work with a sad expression.

"No, Popo, I have never been out there. It is a long walk, and my mamma never lets me go that far from home by myself."

"Mamma Anna would not let me go alone either," Popo said. "But Papa Jean has promised to take Fifina and me soon. I wish you could go with us."

Marcel's eyes sparkled a moment, then they became dark again.

"I'd love to," he said. "I'd love it better than anything, but since I'm working in the shop here I may not have a chance to get away. You see, I'm going to take first communion next Sunday, so what chance have I?"

Popo's eyes opened wide.

"First communion!" he whispered. "We were planning on Sunday; but if you are going to take your communion, I'd rather stay home so as to see you in the procession."

Popo knew that he could not hope to be in a first communion procession. He did not have a pair of shoes. Papa Jean was too poor to buy him any. And of course, no boy or girl would ever take first communion without shoes. But just the same he did not intend to have his trip to the lighthouse spoiled by the fact that Marcel was going to be confirmed.

"Listen," he said after a long pause. "Wouldn't Uncle Jacques let us go on a week day—if we worked very hard till then?"

"Maybe." Marcel gave a little shrug of his shoulders to indicate that he was not at all certain.

"Ask him," Popo urged.

"You come with me," Marcel said. "He might not be so apt to say no if both of us ask him."

"All right."

Uncle Jacques was bent over a low bench, with his knee on a strip of board to hold it rigidly in place. His shoulders were rising and falling with a regular motion, for he was sawing and he seemed to be deeply absorbed in keeping his line straight.

The boys waited till he had finished. Seeing them at his elbow, Uncle Jacques looked up.

"Well?" he said, half surprised.

Marcel lost his courage. "You tell him, Popo."

"Papa Jean promised to take Fifina and me to the lighthouse next Sunday, and I wanted Marcel to come. But he is to take first communion then. Couldn't we go on a week day if we work well till then?"

Uncle Jacques rested his saw on the bench and put his hands into his pockets. He walked to the door and looked out at the quiet street.

"To the lighthouse," he mused.

"We have never been," Popo urged.

"No, we have never been," Marcel repeated.

"Are you sure my brother Jean could take you on a week day? You know he fishes every day except Sunday, the Lord's Day. I wonder if he'd want to lose a day."

Popo could not answer. He stood digging his big toe into the dirt floor of the woodworking shop. Marcel showed his disappointment. Uncle Jacques looked down at the youngsters, and after a moment or two his lips parted in a smile.

"There now!" he said. "Don't feel disappointed so soon. I'd like you boys to have a trip to the lighthouse. It is worth seeing. In fact, I have just been thinking that if Jean is willing we might all take a trip out there. He could stay home from his fishing, I could leave the shop to old man Durand, and then every one could go— your Mamma Anna, my wife, every one. A regular picnic. The womenfolk might fix something to eat."

Both boys looked up happily.

"I am sure Papa Jean will be glad to go," Popo said. "He promised to take us on the Lord's Day, and when he finds that Marcel will be in the procession, I know he will want us to stay home to see the sight."

— 83 —

"Well, you ask him to make sure," Uncle Jacques said, picking up his saw again. "Day after to-morrow will suit me all right."

When Popo went home that evening, he immediately told the glorious prospect to Fifina.

"Oh, Fifina," he fairly sang. "I've got a thing to tell you that'll burn your ears. I've got a thing to tell that will make you laugh and cry at once, but part of it isn't known yet."

"How is that, *petit monde?* What do you mean by that riddle, little world?"

"Uncle Jacques will take a day off if Papa will do the same. Uncle Jacques will take his family for a trip to the lighthouse day after to-morrow if Papa Jean will do the same. We can have a picnic if Papa Jean is willing to have it on a week day instead of Sunday."

"Let's tell Mamma Anna."

"Yes, let's do."

"Oh, Mamma Anna," they sang together. "I've got a thing to tell that will burn your ears, burn your ears."

"Tell me, *petit monde*, tell me."

"Uncle Jacques will take his family for a trip to the lighthouse day after to-morrow if Papa Jean will do the same. We can have a picnic if Papa Jean is willing."

"There he is on the beach," said Mamma Anna. "Now he is coming up the slope. Let us ask him."

"Oh, yes, let's do, let's do!"

Papa Jean came into the yard and stood outside the door with his string of glistening silver-green fish. He was tired but apparently happy. His shoulders were wet with perspiration and shiny like metal.

"Oh, Papa Jean! We've got a thing to tell that will burn your ears, burn your ears, burn your ears," the three sang.

"It's a riddle, little world. Tell me, tell me."

"Not exactly a riddle, Papa Jean. Uncle Jacques will take a day off if you will do the same. Uncle Jacques will take his family to the lighthouse day after to-morrow if you will do the same. We can have a picnic if you are willing."

When they had calmed down, Papa Jean said:

"Of course I am willing. I promised you that. But why not go on Sunday as we first planned? Why must I lose a day? I am a poor man."

"Marcel is going to take first communion next Sunday, Papa Jean. He could not come; Uncle Jacques's family could not come. And besides, we want to see Marcel march in the procession to the church."

"Oh, I see."

Papa Jean bowed his head. Popo watched him closely. Above the man's shoulders the youngster could see the bright colors of the sunset.

"Say yes, Papa Jean. Please say yes," Popo urged.

"Please say yes," Fifina said.

"Please say yes," Mamma Anna whispered.

When Papa Jean looked up, he was smiling pleasantly.

"Yes," he said softly. "We will go."

XIII

A GRAND TRIP

THE day for the picnic came. Popo got up early and went to the fountain for a pail of water. Fifina took care of Pensia while Mamma Anna prepared a lunch, and Papa Jean walked around the house smoking his pipe. The sky was clear and blue, but there were a few rosy clouds near the eastern horizon. Popo skipped for joy on the way to the fountain. When he had filled his pail, he placed it carefully on his head, balanced it so that it would not fall, and started home. He could carry a pail of water on his head quite easily now that he was becoming a big boy. He had practiced, and now it was not hard to do. Being so happy about the picnic made it seem easier.

When Popo got home, Uncle Jacques, Aunt Melanie, and Marcel were sitting on a bench in the yard. They had brought a large basket of mangoes and bananas and were ready to start the journey.

"Hello, cousin," Marcel called as Popo came into the yard. "When did you learn to carry water on your head?"

"A long time ago," Popo smiled modestly. "Mamma Anna taught me."

"Yes," Mamma Anna joined in, "now that he is getting to be a big boy, he must learn to do things."

"You are right," Aunt Melanie said sadly. "Marcel could learn a lesson from Popo. I like to see boys help their mothers."

Uncle Jacques and Papa Jean walked down to the water's edge while they waited. Soon, however, Mamma Anna called to let them know that she was ready, and immediately they started up the slope. As they walked, they talked together in quiet, subdued voices.

"So we're all ready!" Papa Jean said.

"Yes," said Mamma Anna. "You men will have to carry the lunch baskets. I shall have Pensia in my arms, and Aunt Melanie is not feeling so very strong. She had better not try to carry a load."

"Well, all right," the men agreed. "Come along then, we don't want you weaklings to lag behind." They laughed indulgently.

Uncle Jacques and Papa Jean led the way, still talking quietly. They turned into rue Bord de la Mer in the direction that led out of town and toward the entrance to the harbor of Cape Haiti. Popo and Marcel and Fifina came scampering behind them, while Mamma Anna and Aunt Melanie brought up the rear.

They walked steadily for about a mile above the sea, passing houses, trees, and hills on one side. On the other they passed a number of boats tied up to the shore, and nets stretched out to dry on the low trees. There were several small beaches too. One of them belonged to the American Marines. Their bathhouses were built near the water, and the water itself was fenced in to keep the sharks away from the place where the Marines bathed.

Beyond this were the beaches where the people of Haiti bathed. These also were protected from sharks—but not by wire fences built in the water. They were protected by natural reefs, and the seaweed, that was so thick a short distance out from the shore, could be clearly seen from the hillside path as Papa Jean and Uncle Jacques with their families began the descent to the ocean's edge.

In one place the children stopped to look at a group of teachers from the Catholic school in Cape Haiti as they bathed in the surf. The spot which they had selected was very rocky. The mother superior sat on a stone on a little hill away from the water. She shielded the teachers from view and helped them while they changed their black robes for white bathing gowns. Then she let their hair down for them. These women all white-clad and looking very much like angels, walked slowly and solemnly into the water and bathed sedately among the big rocks that jutted up out of the sea. It was a beautiful sight. The water was still and blue-green. Where it washed against the stones, it threw up a bit of white foam. The teachers went out where the water was deep enough to reach their shoulders. Their hair floated on the water. And all the while the mother superior watched them lovingly from her rock on the hill.

Popo and the others walked a little farther and then set their lunch baskets on the sand.

"What shall we do first?" asked Uncle Jacques.

"Let's stay here till after dinner," said Mamma Anna. "It will be better to bathe now and then go to the lighthouse after eating. You know, it is not good to swim right after a meal."

"Yes," said Aunt Melanie. "That is the best way."

The plan pleased the children. They were all anxious to see the lighthouse, but they were also eager to bathe in the ocean; and they did not mind putting off the visit to the lighthouse for a few hours. They quickly squirmed out of their clothes—none of them wore more than two pieces—and dashed out into the surf. Uncle Jacques and Papa Jean also got ready for a swim. Mamma Anna and Aunt Melanie sat on the sand near a flat rock and played with Baby Pensia.

All the children were at home in the water. They dived in and easily swam out to the reefs. Then they returned to where the water was not too deep and began playing in the seaweed. They found long beautiful switches and garlands of green weed and drew them out of the water. But they soon saw that the seaweed was not nearly so pretty out of the water. It felt slimy and was a dull brown color when held up to the sun.

"Leave it where it ought to be—in the water," said Mamma Anna.

Fifina came up on the sand to rest. A little later Popo and Marcel heard her calling.

"Oh, see what I have found, Popo! Come and see, Marcel!"

She had found a fine sea shell, rosy and pink inside, and almost the color of silver on the outside.

"That *is* a beauty," Marcel said.

"Yes, indeed," said Popo.

Papa Jean and Uncle Jacques were swimming along the far reef with long powerful strokes. Two or three hours passed, and the sun climbed the sky steadily. In a short time, it was overhead.

Mamma Anna and Aunt Melanie spread the baskets of lunch on a flat rock and called their children and husbands. The children came leaping out of the water like young animals, but the two brothers took their time. They swam reluctantly to the shore and walked slowly up the beach.

The food tasted good after so much exercise. Popo got his mouth smeared and sticky with the sweet juice of the mangoes. He licked his fingers. It seemed to him that nothing had ever tasted so fine before. Fifina liked bananas best, but Mamma Anna would not let her eat more than three. Too many bananas might make a small child sick.

After dinner the men and the children put on their clothes and started for the lighthouse. The women stayed behind, playing on the sand with Pensia.

The lighthouse path led up a hill and under the overhanging branches of almond trees, tropical oaks, and the long dangerous arms of giant cactus. Occasionally it came out of the shadows, and Popo could see the water farther and farther below, at the foot of high steep cliffs. The waves looked smaller, and the ocean seemed smoother. But he did not have much time to notice the view. The lighthouse was still a good distance ahead, and Papa Jean and Uncle Jacques were walking steadily.

Suddenly Uncle Jacques, who was leading, called over his shoulder: "Look out for the snake. Don't hurt it."

That sounded very strange. Most people would have said, "Don't let the snake hurt *you*." But Popo knew exactly what he meant. A moment later he saw the thin green thing, beautiful and delicate, hanging across the path like a twig. Its head was in the bushes on one side of the path, and its tail was in the bushes on the other side. Popo had always heard his father say that the slender green snakes would not hurt any one. He had also been told that it was wrong to kill them. They had for years been regarded as sacred by the people of Haiti.

The little group passed through the ruins of two or three abandoned forts, in secluded parts of the mountain side. They had been built long ago, Uncle Jacques said, when Haiti was still a French possession. Old rusted guns pointed up

out of the crumbled stonework. These forts reminded Popo of the sad story about the Citadel.

"Well, here we are!" Uncle Jacques cried.

Popo had been so busy thinking that he was amazed to see the lighthouse just a few steps ahead. And what a view! The great Atlantic!

The lighthouse was a tall, round white structure with big lenses near the top, like huge eyes.

"That is where the light comes out at night," said Uncle Jacques. "When everything is dark, that light flashes far across the water to warn passing ships of the dangerous rocks on our coast. These ships carry grains, meats, sweet woods, ornaments, and all manner of treasures. They go to France and Germany and England and the United States, and they must not be wrecked on our rocky coast. Sometimes ships come to us bringing cloth, shoes, canned foods, and wine. They must not be lost, either. They must bear their cargoes safely. When they arrive in the middle of the night and see our lighthouse, they know that they have reached their destination.

So they drop anchor and wait till morning to be piloted into port by our own men who understand the passages and the reefs."

"It's a wonderful light," Popo said.

"Indeed," said Marcel.

"How beautiful the sea is from here!" said Fifina.

"Yes," said Papa Jean, "but do you see the sky?"

All the others looked up at the same moment. Sure enough, the west was covered with black heavy clouds.

"It's going to rain," said Uncle Jacques.

"H'm. We'd better hurry back," said Papa Jean.

Popo took one more look at the wide sea where the great ships pass. Then all turned and started down the hill quickly to get Mamma Anna, Aunt Melanie, and Pensia before the storm.

The sky became darker and darker, and a wind came up. It went through the tropical trees like music. Birds hid in the thickets and cried. The bathing beaches were deserted, and only Mamma Anna and Aunt Melanie were left on the sand. They had their baskets ready, and Mamma Anna had Pensia in her arms.

When the group came down on the sand again, the first drops fell.

"We must hurry," said Mamma Anna.

"Yes," said Aunt Melanie. "This is the first rain of the season, and I'm afraid it will be a heavy one."

The men took the baskets and started out in the path. The youngsters followed close. Behind them came the mothers with Pensia. There was no loitering this time. Everybody was excited. Everybody was anxious to get home.

"We must hurry," one said.

"We must hurry," another repeated.

That was all Popo heard. That was all Fifina heard. Everybody was saying, "We must hurry."

A few more drops fell, and the sky got blacker and blacker. The wind whistled. It blew through the trees like heavy music, like horns and drums. The poor frightened birds in the thickets cried louder and louder.

Suddenly the storm broke. A peal of thunder rattled the roof of the sky. The clouds were so low that Popo could not see the tops of the mountains. There was a flash of lightning and another clap of thunder. Then the rain came, a great sheet of warm white rain.

The sea became noisy. The waves boomed, and the long green tongue of the water licked the

Everybody was saying, "We must hurry."

white sand. The sky boomed, and water poured down.

The little group walked faster and faster. But Popo was not afraid. He was used to the water, so he did not mind getting wet. But just the same it would be good to get home, to be in the house and to put on a dry shirt.

Back in town the rain was still pouring and the gutters were streaming. Popo and Marcel waded in them, splashing the water with their feet. The two brothers, walking ahead, did not mind the storm either. They had seen many storms. They talked together quietly while the thunder rattled over the hills.

A little farther on, a small goat was standing alone in the street unsheltered from the rain. His back was turned to the wind, and he was drenched. He looked very poor, very helpless there in the heavy downpour all by himself. Popo wondered what he could do to help the poor creature.

"Look at that poor goat," he said to Marcel. "I wish I could help him."

"There are thousands of goats in this town," Marcel laughed. "Would you like to help them all? There are more goats than there are people."

"I'm not making fun," Popo said seriously. "He looks so pitiful out here alone."

"That shed over there is his house," Fifina said. "Why don't you make him go inside?"

"I will," Popo said eagerly.

He ran to the goat and tried to push him

toward the shed. But the goat was blinded by the rain and did not seem to understand that Popo was trying to help him. Suddenly, the confused creature turned around abruptly and butted Popo in the stomach. It was not a very hard butt, but it caught Popo off balance; and over he went into a mud puddle.

When Popo pulled himself out of the mud, he

heard loud laughter. He was surprised and frightened but not hurt.

"That goat doesn't know what's good for him," he said sadly.

"That's the trouble with most goats, Popo," Papa Jean said. "They don't know what's good for them."

The little group started on again, and very soon they were home.

"Come in and stop with us till the rain is over," Mamma Anna told Aunt Melanie.

"Thanks, but we may as well go on," Aunt Melanie said. "We're all wet, anyway."

Papa Jean stood a long time in the door watching the downpour. He seemed to enjoy the rain. Then, suddenly, he said:

"Popo, this is the beginning of our long rainy season."

"Yes," said Popo. "That's what Aunt Melanie said."

"There won't be much kite-flying from now on, son. You will have to spend your spare time around the house with Fifina and Mamma Anna and Pensia, if you're at the shop."

"I am willing to stay when it rains," Popo said.

"That's a good boy," said Papa Jean. "This summer perhaps you will go with me sometimes in the fishing boat. You will be big enough by then."

Popo was happy to hear this. Fifina too was happy. She thought Popo deserved to go out in the fishing boat and fish like the men.

"That will be fine," she whispered in his ear.

But the summer seemed a long way off. Popo thought that he would indeed be a big boy by that time.

Outside, the rain fell harder and harder. The gutters overflowed. The streets were flooded. A little water began to drip into their house, *tap, tap, tap* on the floor, but Mamma Anna put a pan under the leak.

Popo could not see the beach, the rain was so heavy. He could not even see the tree with the gnarled roots a few hundred yards away. The whole world was wrapped in rain, but the thunder had ceased.

In a little while the storm passed, the sky lightened, and there came a rainbow over the ocean. The world brightened. Fifina put all their wet clothes outside to dry, Papa Jean went off down the street, and Mamma Anna began cooking supper.

"We certainly had a grand trip to-day, even if it did rain," Popo said.

"Yes, indeed," answered Fifina and Mamma Anna together. And even Baby Pensia seemed to agree. "We had a grand trip."

AFTERWORD

ARNOLD RAMPERSAD

Popo and Fifina was the first fruit in publishing of a significant artistic and personal friendship, that between Langston Hughes and Arna Bontemps. Although they seldom lived in the same place, these two writers, who had known each other since 1924, kept resolutely in touch from the start of the collaboration in 1932 that produced this book until Hughes died in 1967. Following *Popo and Fifina,* they worked together on a number of projects, most notably, perhaps, the landmark anthology *Poetry of the Negro* (1949). Their correspondence, a treasure trove of literary and cultural references, amounted to some 2,500 letters. (A selection of this correspondence, edited by Charles Nichols, was published in 1980.)

Hughes and Bontemps brought important differences to their collaboration on *Popo and Fifina.* In 1932, Langston Hughes was almost as close to joining the Communist party as he would ever come (although he never did so). The collapse of his close relationship with his main patron of the 1920s, the

101

wealthy, erratic Charlotte Osgood Mason, in conjunction with the deepening of the depression, had driven him toward radical socialism in the preceding months. This radical turn had also sent him to Haiti, where he lived for three months between April and July 1931. Not long after he returned to the United States, Hughes embarked on a projected year-long tour of the country, starting in the South. His aim was to read his poetry, mainly to black audiences, in churches and school halls or wherever he could. At the end of the tour, in June 1932, he left the United States with a band of African Americans recruited to act in a proposed Soviet film about race relations in the United States. Hughes stayed in the Soviet Union for a year.

Bontemps's life had taken an equally dramatic turn, although by no means to the left. Reared in the Seventh-Day Adventist faith, he had taught for several years at the church-run Harlem Academy in Manhattan, where he had met his wife, Alberta. However, his growing success as a star of the Harlem Renaissance, with its secular, often irreverent concerns, had not sat well with his church superiors. To them, the title of his first novel, *God Sends Sunday*, had seemed like sacrilege, and in 1931 Bontemps was virtually banished, with his family, to another church school, Oakwood Junior College, in Huntsville, Alabama. There, he met even more reactionary attitudes on the part of the school leaders than those he had left in Harlem. Always a man concerned with questions of spirituality, Bontemps was beginning to respond to the political implications of the depression, on the one hand, and the raw racism of the South, on the other, when he started his collaboration with Hughes on *Popo and Fifina.*

The fact that *Popo and Fifina* was a children's book, and one to be published by a major New York City house, meant that both men would have to camouflage or even suppress some of their main concerns if their book were to be successful—or even to be published. Apparently, Hughes and Bontemps accepted this reality. Both men understood the usual requirement that children's books avoid controversy, and both men needed money badly. Macmillan, which had brought out Bontemps's *God Sends Sunday,* offered each an advance of $150 on delivery of the manuscript. Without a hint of trouble, the manuscript was accepted by Macmillan, and it was published in mid-1932.

In bending and adapting his own interests and experiences to write this story, Hughes probably had more compromises to make than Bontemps did. Not only was he far more radical than Bontemps; Hughes had seen firsthand the conditions in Haiti and had written scathingly about them in one of the major communist organs in the United States, *New Masses.* The poverty of Haiti had appalled Hughes. U.S. Marines had been occupying the country for almost 20 years (under a treaty never ratified by the U.S. Senate), but Hughes had observed little that was positive as a result of this repudiation of Haitian sovereignty and autonomy. He had come to despise the Haitian elite, whom he characterized as mainly an effete, snobbish group of mulattoes, content to hold lucrative government sinecures, send their children to school in Europe, engage in petty party strife, ape the manners of Europeans, and relentlessly keep the masses of black Haitians in a state of poverty and powerlessness.

In an essay called "People without Shoes," published in *New Masses,* Hughes vented his anger at conditions in the country. While Haitians of the upper class flourish and the marines live in comfort, "the people without shoes cannot read or write. Most of them have never seen a movie, have never seen a train. They live in thatched huts or rickety houses; rise with the sun; sleep with the dark. They wash their clothes in running streams with lathery weeds—too poor to buy soap. They move slowly, appear lazy because of generations of undernourishment and constant lack of incentive to ambition.... They grow old and die.... The rulers of the land never miss them. More black infants are born to grow up and work. Foreign ships continue to come into Haitian harbors, dump goods, and sail away with the products of black labor."

Hughes and Bontemps wanted to write a story precisely about the "people without shoes" and from the point of view of those same people. From the start of his poetic career, Hughes had placed an identification with the masses of blacks at the center of his consciousness as an artist. He had made the blues, the music of the black American masses, pivotal to the evolution of his art as a poet in the 1920s. His political radicalism had also been reflected as one major strain in his writing from the start of his career, although he had also written extensively as a lyric poet without reference to race. Bontemps had been less overtly political in his aesthetic, but he, too, was a man committed to a democratic view of African Americans. Certainly, he would have had no difficulty in seeing Haiti through Hughes's eyes and in helping to write a story that the more radical Hughes would find acceptable, given their intended audience and their publisher.

Popo and Fifina also reflects Hughes's active interest in the Caribbean and the interest of both men, perhaps to differing degrees, in the concept of Pan-Africanism. Before visiting Haiti, Hughes had been to Cuba three times, and he had lived in Mexico twice, including one stay of a year. His sojourns in Mexico had come before he was 20, but his visits to Cuba were more recent and had both reflected and stimulated his radicalism. He had been fascinated by black culture in Cuba, and he had visited Haiti precisely because it was a predominantly black nation. For Hughes, Haiti was not only mainly black but also of unrivaled historical and symbolic value as the only independent black republic (despite the presence of the U.S. Marines) in the Western Hemisphere. Hughes was well aware of the history of Haiti's successful struggle against European imperialism, the defeat of the French army, and the establishment of the republic. He had made a point of visiting, on three separate occasions, the ruins of the mighty citadel built by the Haitians as the ultimate defense against European armies.

Pan-Africanism, a concept that emphasizes the unity of people of African descent around the world, appealed to Hughes (and no doubt also to Bontemps) despite his undeniably cosmopolitan interests. Pan-African Congress meetings, organized mainly by the outstanding black American scholar and intellectual W. E. B. Du Bois (for whom both Hughes and Bontemps had great respect), had taken place several times since the end of the World War in 1918. However, Hughes and Bontemps were probably much more informal than Du Bois in their interest in Pan-Africanism. Like many other racially conscious young writers of their time, they felt a sentimental, if urgent, interest in black people in other lands and were less easily captured by the geopolitical aspects of the matter, involving colonialism and imperialism, that seemed to fascinate Du Bois. *Popo and Fifina* would reflect this vaguer, perhaps more sympathetic and sentimental feeling.

Many of Hughes's radical ideas about Haiti are simply unexpressed in this book. There is no attack on the mulatto leadership class—indeed, no mention of this class at all. Whereas Hughes had complained in his polemical writings about swaggering white soldiers insensitive to local feelings, and about the failure of social services under the control of the U.S. Marines, in *Popo and Fifina* there is only a single reference to the foreign soldiers.

Hughes and Bontemps mention the existence of a marine-controlled beach, which is protected from attacks by sharks by a sort of fence planted in the water. (The beaches for locals are protected from sharks by nature, mainly in the form of reefs.) In this story, we see endless examples of poverty (in a country where the average daily wage, as Hughes revealed elsewhere, was 30 cents) but virtually no pictures of affluence. The result is that the reader exists solidly within the world of the poor, but without the sense of outrage that comparisons to wealth inevitably would have brought.

Instead of providing conflicting illustrations of wealth and power or of the exploitation of destitute blacks, Hughes and Bontemps concentrate on showing the simple, ordered, industrious, resourceful lives of the typical Haitian poor. These poor people must be typical, because we are never invited to see Popo and Fifina, and their family, as anything else. In a gorgeous tropical land of bright flowers, birds, and fish, of succulent fruit and cool mountain streams, the masses of people lead exemplary lives, making the most of what they have. Family and kinship are central factors in creating stability. In carefully defined roles (old-fashioned and patriarchal, to be sure), the members of the family together form a solid unity. A larger sense of kinship enables Popo to become an apprentice as a carver of wood. Beyond family, in addition, a spirit of cooperation among the poor also obviously exists, so that Popo's father can immediately find a place among the fishermen when he moves to town.

No threat of violent crime, or any other kind of crime, appears to mar this community. A sense of the necessity of dignity, tradition, and order rules. The influence of religion is benign, even though we are exposed to the celebrated spiritual dualism of Christianity and paganism in Haiti. We are given a passing glimpse of the "pagan" religion of the black poor, but this glimpse is through Popo's boyish eyes, so that we see little. And we also learn respectfully, if also fleetingly, about the first communion of Marcel, Popo's cousin. The black poor of Haiti, absorbing both traditions, appear to benefit from both. In real life (as Hughes reported elsewhere), the marines and the Catholic church frown on drumming, for fear of insurrections, and the largest drum (there are three basic sizes) is even forbidden; but in *Popo and Fifina* the largest drum is beaten boldly, as it often was in clandestine defiance of the powers, at the nocturnal meeting attended by Popo.

This picture of the Haitian poor is, of course, not inconsistent with a radical socialist view of the masses. However, Hughes and Bontemps make no attempt to mine this connection openly, to write a drama in which the virtues of the people find a discernible political focus. Nevertheless, they may have carefully hidden at least one revolutionary image in the text. When Papa Jean makes a kite for Popo, it is red and shaped like a star—the symbol of the Red Army of the Soviet Union. This red star, soaring triumphantly in Popo's hands, battles an aggressor "hawk" kite and defeats it.

In all other ways, *Popo and Fifina* is a gentle, episodic narrative, a way of exposing and explaining to us a few of the customs and traditions of the Haitian masses. The story ends with a picnic excursion to a lighthouse, completed during the first downpour of the rainy season. Thus, although the sun comes out and a rainbow appears, something like a veil of mist is drawn across Haiti in the end, and the reader withdraws quietly after being privileged to share in the lives of the people, mainly in the company of a little boy and his sister. It is important to see, however, that by painting such a dignified, moral picture of black Haitians, Hughes and Bontemps were making—in the context of their time—a bold statement about race and culture.

Popo and Fifina was well received both by critics and the buying public when it appeared in 1932. Although it was never a best-seller, the book stayed on the Macmillan list, generating a steady though hardly spectacular yield in royalties for Hughes and Bontemps, through the following generation. Finally, as Hughes saw it, the book was dropped by the press simply to make room for another title. (In 1953, however, after years of being attacked as a leftist, Hughes was subpoenaed by Senator Joseph McCarthy and questioned about his communist connections. Although he was exonerated by McCarthy, it is possible that the anticommunist feeling of these years was enough for Macmillan to move to drop him from its list of authors.)

Both Hughes and Bontemps went on from *Popo and Fifina* to make the writing of children's books an important part of their careers, although aside from his collection of poems, *The Dream Keeper* (1934), Hughes's next book for children was not published until 1952. Both men liked writing such material, and as professional writers they both also saw the children's book as a convenient way of earning money without committing oneself to a grand project—for which both men, as blacks, would never have received (certainly not before the 1960s) the large advances that sustained many other writers whose literary talent and publishing records were inferior.

Hughes eventually published about a dozen children's books altogether, for a variety of ages. He wrote several *First Book* volumes for the publisher Franklin Watts, starting in 1952 with his *First Book of Negroes*, in addition to books for older children, such as *Famous Negro Music Makers* (1955), for Dodd, Mead. Bontemps published ten children's books, including three with Jack Conroy, whom he met in Chicago, where Bontemps lived for several years; his works include both fiction and biographies of Frederick Douglass and George Washington Carver. After *Popo and Fifina*, Hughes and Bontemps wrote several stories for children. Although none of these stories was published, they testify both to the rich talent of Hughes and Bontemps as artists and to their ability to inspire one another, especially when their primary audience was the young reader.

Langston Hughes (1902–1967) was born in Joplin, Missouri, and grew up in Kansas and Ohio. He moved to New York City when he was 19 years old to attend Columbia University. He was one of the most versatile writers of the artistic movement known as the Harlem Renaissance. Though known primarily as a poet, Hughes also wrote plays, essays, novels, and a series of short stories that featured a black Everyman named Jesse B. Semple. *Popo and Fifina* was his first novel written especially for children and his first published collaboration with fellow poet Arna Bontemps.

Arna Bontemps (1902–1973) was born in Louisiana and grew up in California. He moved to New York City in 1923, and it was there that he met Langston Hughes and other writers who were leaders of the Harlem Renaissance. Bontemps began his literary career as a poet but also wrote novels and edited anthologies of African-American poetry and folktales. He worked with Hughes on several novels for young adults and several anthologies, such as *Poetry of the Negro, 1746–1949* and *The Book of Negro Folklore*.

E. Simms Campbell (1906–1971) was an illustrator and cartoonist for *Esquire, The New Yorker, Life, Collier's, The Saturday Evening Post,* King Features Syndicate, and *Opportunity,* the publication of the National Urban League. Born in St. Louis, he graduated from the Art Institute of Chicago and held honorary degrees from Lincoln and Wilberforce Universities.

Arnold Rampersad is professor of English and director of the American studies program at Princeton University. His books include *The Art and Imagination of W. E. B. Du Bois* and the *The Life of Langston Hughes* (2 volumes).

THE IONA AND PETER OPIE LIBRARY OF CHILDREN'S LITERATURE

The Opie Library brings to a new generation an exceptional selection of children's literature, ranging from facsimiles and new editions of classic works to lost or forgotten treasures—some never before published—by eminent authors and illustrators. The series honors Iona and Peter Opie, the distinguished scholars and collectors of children's literature, continuing their lifelong mission to seek out and preserve the very best books for children.

jHUGHES   38841

Popo and Fifina

DATE DUE

JUL 28 '98			
AUG 11 '98			